What Are Rocks and Minerals?

Copyright © by Harcourt, Inc.

All rights reserved. No part of this publication may be reproduced or transmitted in any form or by any means, electronic or mechanical, including photocopy, recording, or any information storage and retrieval system, without permission in writing from the publisher.

Requests for permission to make copies of any part of the work should be addressed to School Permissions and Copyrights, Harcourt, Inc., 6277 Sea Harbor Drive, Orlando, Florida 32887-6777. Fax: 407-345-2418.

HARCOURT and the Harcourt Logo are trademarks of Harcourt, Inc., registered in the United States of America and/or other jurisdictions.

Printed in Mexico

ISBN 978-0-15-362228-1
ISBN 0-15-362228-8

2 3 4 5 6 7 8 9 10   050   16 15 14 13 12 11 10 09 08

Visit *The Learning Site!*
www.harcourtschool.com

Let's Rock

You see rocks every day, inside and outside. If your classroom has a chalkboard, chances are the chalk itself is made from a kind of rock.

Outside, we also see rocks. As we ride down the road, we may see buildings, fences, or signs made from rocks and stones. Even the streets and roads we ride on are made with materials from rock.

A rock is a natural material. It is made up of one or more minerals. A **mineral** is a solid substance that occurs naturally in rocks or in the ground. There are over 4,000 minerals. Each one is different.

Rocks and stone are often used to add to the beauty of a building.

Diamond is the hardest mineral known. Talc is the softest mineral. We get talcum powder from talc.

diamond

talc

There are a lot of ways to tell the difference between minerals. Minerals are different colors. They can be shiny, dull, hard, soft, or even magnetic. They reflect light. The way a mineral reflects light is called its *luster*.

Many minerals look alike. Scientists use the physical properties of minerals to tell them apart. For example, scientists can compare the hardness of two minerals. A mineral called moissanite looks a lot like diamond, but moissanite is softer.

Scientists can also tell the difference between minerals by their streak. When rubbed across an unglazed tile, a mineral leaves a streak of powder. Scientists use the streak's color to identify the mineral.

MAIN IDEA AND DETAILS Name at least two ways scientists can tell the difference between minerals.

Types of Rocks

Did you know there are three different types of rocks? **Igneous** rocks form when melted rock cools and hardens. Some places beneath Earth's crust are so hot that rock there is liquid. This rock is called **magma**.

When volcanoes erupt, magma comes to Earth's surface. Magma that comes to Earth's surface is called **lava**. Lava cools and hardens into igneous rock much quicker than magma.

Sedimentary rocks are formed by pieces of rock that are broken down and carried by water, ice, and wind. The pieces, or sediment, settle on Earth's surface and press together in layers.

Sometimes you can tell which group a rock belongs to by its appearance.

igneous rock: granite

metamorphic rock: slate

To *morph* means to "change." **Metamorphic** rocks are rocks that have changed as a result of high temperature and pressure. Metamorphic rocks may once have been igneous, sedimentary, or even other metamorphic rocks. Slate is a metamorphic rock that began as shale. Increased temperature and pressure caused it to change.

> **Fast Fact**
> Quartz can be found in many places. It is used for making glass, watches, and computer chips. A form of it can also be used to help make hard surfaces smooth.

Rocks are grouped based on how they form. Igneous rocks, including granite, are usually shiny and hard. Sedimentary rocks like limestone, chalk, and shale are soft and have layers. Fossils are common in sedimentary rock.

Metamorphic rocks can have bands. The flattened grains and curves in metamorphic rocks are evidence of the pressure that formed them. Marble and slate are metamorphic rocks.

 MAIN IDEA AND DETAILS Describe how the three different groups of rocks form.

sedimentary rock: sandstone

The Rock Cycle

The **rock cycle** describes the processes that change rocks. The rock cycle takes many, many years.

The rock cycle includes temperature changes, pressure, and movements that occur in nature all the time. Volcanic eruptions change rocks, and earthquakes move them. The powerful rain and wind on mountains also change rocks. Water movement in rivers and oceans carries sediment to new places. New rocks are formed where the sediment is deposited.

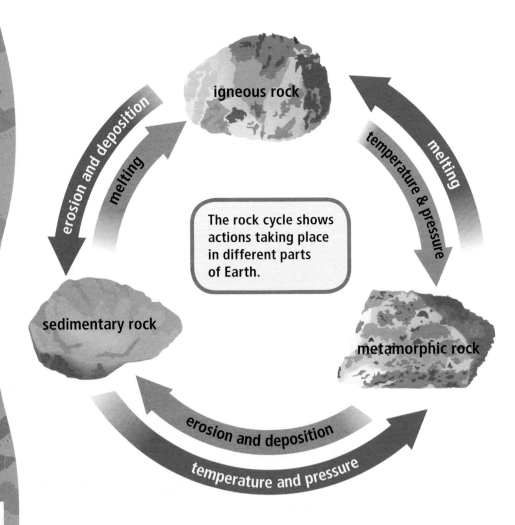

The rock cycle shows actions taking place in different parts of Earth.

Lava can reach temperatures of 1200°C and higher.

Fast Fact

Earthquakes move rocks, but they are sometimes too small to be felt by humans. From 1975–1995, every U.S. state other than Florida, Iowa, North Dakota, and Wisconsin had an earthquake. Most were not felt!

The effects of nature can cause rocks to change into other kinds of rocks. Suppose an earthquake causes an igneous rock to split and roll into a river. As the rock tumbles down the river, it breaks into pieces. The pieces flow into the ocean with other small pieces of rock. As more sediment collects, the weight presses the pieces together into sedimentary rock.

If rock is pushed until it goes deep underground where the pressure and temperature increase, the rock becomes a metamorphic rock. It could eventually melt into magma. Then a volcano could erupt. The magma flows out as hot lava. The lava cools on Earth's surface and hardens into an igneous rock. This change can happen over millions of years as part of the rock cycle.

SEQUENCE If a piece of rock is sitting on the ocean floor, what has to happen for it to become sedimentary rock?

What Is Weathering?

Weathering is the breaking down of rocks on Earth's surface into smaller pieces. Weathering also changes the shape of rocks and the surface of Earth. Soil is created partly as a result of weathering.

One kind of weathering is called *mechanical weathering*. Mechanical weathering causes rocks to break into smaller pieces. For instance, drastic changes from hot to cold or cold to hot create cracks in a rock. Water can fill the cracks and then freeze. When the water freezes, it expands and causes the crack to open more. The rock finally breaks into pieces.

Ice expanding in the cracks of this rock caused it to break into pieces.

It is clear from the pits in this rock that weathering has occurred.

Wind and water also cause mechanical weathering by pounding on rocks. Even trees cause weathering. As a tree root grows larger in the crack of a rock, the rock splits. This is sometimes called biological weathering.

Another kind of weathering is chemical weathering. Did you know a rock could rust? Water mixes with iron in the rock. The iron changes into other materials such as rust. In other cases, water mixes with chemicals or minerals. As the water moves, it causes rocks such as limestone to dissolve and sometimes completely break down.

It is not hard to find the effects of chemical and mechanical weathering on rocks. Sometimes a rock's corners are rounded. Sometimes there are dents in a rock's surface. Along a mountainside, you might even see that a layer of rock has peeled away.

 CAUSE AND EFFECT What is the result of mechanical weathering?

Endless Erosion

Erosion is the process of moving sediment by wind, water, or ice from one place to another. Like weathering, erosion affects rock on Earth's surface.

The Grand Canyon is an excellent example of water erosion at work. The Grand Canyon was formed by the Colorado River. Over millions of years the river has worn away and moved pieces of rock. Then wind, rock falls, and mudslides made the canyon even wider.

Some scientists suggest that the canyon was formed by a large amount of water rushing through over a short period of time. In either case, erosion was at work.

Erosion has made the Grand Canyon one of our most spectacular national parks.

Alaska's Hubbard Glacier is one of the largest glaciers in North America.

In the desert, wind is a strong erosion force. As sand and stones blow against rock, the rock wears away. The wind carries the small pieces of sand and rock from place to place. Sand dunes are built up. Over time the sand dunes can become mountains of sandstone.

Glaciers are another natural cause of erosion. Glaciers are huge sheets of ice. At times glaciers have covered much of Earth's surface. These periods are known as ice ages.

As glaciers move over land they scrape the ground. As this happens, they move huge rocks and boulders as well as small rocks and sediment.

Fast Fact

In 2002, Hubbard Glacier created a natural dam when it moved rocks and sediment and blocked the flow of water from Alaska's Russell Lake. Luckily, the trapped water finally broke free in August of that year.

CAUSE AND EFFECT Explain how wind can cause erosion.

Soil Under Our Feet

Over many years of weathering, rocks break down into fine pieces called sediment. The sediment mixes with **humus**, the remains of decayed plants and animals, and forms soil. Soil also contains water and air. These provide important nutrients for plants and seeds.

Soil is a rich environment for living things. It provides food for plants. Plants, in turn, serve as food for animals and people. But soil is not only a source of food. It is also a home for living things such as insects and bacteria.

It takes a long time for soil to form. Therefore, soil has to be protected in order to be useful for a long time.

An earthworm is one of the many living things that make their homes in the soil.

Soil has **horizons**, or layers. Each horizon has particles that are different sizes. The upper layers are near Earth's surface. These layers are often very rich because they usually contain humus.

The next layer is underground. It is not as rich, because there are not as many decayed plants and animals. This layer may contain some minerals that have been carried into the ground by rain. It also contains rock that has weathered.

The bottom layer of soil is known as **bedrock**. It is mostly made of solid rock. It lies under the other layers.

 COMPARE AND CONTRAST Compare the richness in the soil horizons.

The soil in this garden is rich in plant nutrients. Notice how dark it is.

Types of Soil

Different types of soil have different kinds of soil particles. Sand has the largest particles. Each particle is about 1–2 millimeters across. Sand is rough to the touch, so soil with a lot of sand in it will feel rough.

You probably know that few plants grow in the desert. One of the reasons is that the soil in the desert is mostly sand. Water moves quickly through sand. It flows downward, away from the topsoil. Few plants can survive because there is very little water in the topsoil.

Because sand does not hold water to bring nutrients to plants, not much grows in the desert.

Clay has the smallest soil particles. Each particle is about 1/1,000 the size of a particle of sand. Clay gets sticky when it is wet and becomes hard and smooth when it is dry. Clay-rich soils are not good for plants either, because they do not allow water to drain.

The best type of soil for plant growth is a mixture of sand, clay, and silt. Silt is smaller than sand but bigger than clay. Silt can hold water but still allow it to drain. It can also store nutrients for plants to use.

 COMPARE AND CONTRAST What are the differences in the ability of clay and sand to support plant life?

Clay-rich soil holds too much water. This soil is not good for planting.

Summary

Rocks are made of minerals. The three types of rocks are igneous, sedimentary, and metamorphic. They are formed during stages of the rock cycle. Part of the cycle involves weathering and erosion. As rocks break down, soil is formed. Soil provides nutrients for plants.

Glossary

bedrock (BED•rahk) The solid rock that forms Earth's surface (13)

clay (KLAY) The smallest particles that make up soil (15)

erosion (uh•ROH•zhuhn) The process of moving sediment from one place to another (10, 11, 15)

horizon (huh•RY•zuhn) A layer in the soil (13)

humus (HYOO•muhs) The remains of decayed plants or animals in the soil (12, 13)

igneous (IG•nee•uhs) A type of rock that forms from melted rock that cools and hardens (4, 5, 6, 7, 15)

lava (LAH•vuh) Molten (melted) rock that reaches Earth's surface (4, 7)

magma (MAG•muh) Molten (melted) rock beneath Earth's surface (4, 7)

metamorphic (met•uh•MAWR•fik) A type of rock that forms when heat or pressure change an existing rock (4, 5, 6, 7, 15)

mineral (MIN•er•uhl) A solid substance that occurs naturally in rocks or in the ground (2, 3, 13, 15)

rock cycle (RAHK CY•kuhl) The sequence of processes that change rocks from one type to another over long periods (6, 7, 15)

sedimentary (sed•uh•MEN•ter•ee) A type of rock that forms when layers of sediment are pressed together (4, 5, 6, 7, 15)

weathering (WETH•er•ing) The breaking down of rocks on Earth's surface into smaller pieces (8, 9, 10, 12, 13, 15)